Longman Practice Exam Papers

A-level Physics

John H. Avison

Series editors:

Geoff Black and Stuart Wall

Titles available for A-Level

Biology

Business Studies

Chemistry

Physics

Psychology

Pure Mathematics and Mechanics

Pure Mathematics and Statistics

Addison Wesley Longman Limited
Edinburgh Gate, Harlow
Essex CM20 2JE, England
and Associated Companies throughout the World

First published 1999

ISBN 0-582-36922-3

British Library Cataloguing in Publication Data
A catalogue record for this book is available from the British Library

Set in Times 11/13 and Gill Sans by 38

Printed in Singapore through Addison Wesley Longman China Ltd, Hong Kong

Contents

Editors' preface

Longman Practice Exam Papers are written by experienced A-level examiners and teachers. They will provide you with an ideal opportunity to practise under exam-type conditions before your actual school or college mocks or before the A-level examination itself. As well as becoming familiar with the vital skill of pacing yourself through a whole exam paper, you can check your answers against examiner solutions and mark schemes to assess the level you have reached.

Longman Practice Exam Papers can be used alongside *Longman A-level Study Guides* and *Longman Exam Practice Kits* to provide a comprehensive range of home study support as you prepare to take your A-level in each subject covered.

Introduction

What happens in real examinations?

- In real examinations the actual grade boundary marks vary from year to year and from board to board as the panels of examiners make judgements about the standards and quality of answers they see before them each year.
- The grade A boundary is determined separately. The boundaries between grades E and B are divided, giving equal marks ranges between all grades.
- When a particular exam paper turns out to be 'easier' than usual, the examiners set the grade boundaries at higher marks, so there are no 'standard' of fixed 'pass' marks or grade A marks.
- Individual papers and whole examinations are not usually marked out of 100. The total number of marks relates to the number of identified marking points which you, the candidate, are expected to earn in the time allowed for the examination.
- It helps to remember that the examiners have an agreed mark scheme; that is, a list of specific points which they are looking for in your answer. The number of marks indicated at the end of each section of a question gives you a clear idea about how many points are required and how long or detailed your answer should be. It is important to look at the mark allocation before starting your answer.
- Some panels of examiners list more marks in their marking schemes than are allocated to the question on the paper. This means that you can earn full marks without actually getting all the points on the mark scheme, or even without getting the question completely right.

How to avoid throwing marks away in a Physics examination

Marks are often lost through careless answers or through not understanding the 'rules' by which you should engage in the 'contest' with the examiners. Calculations, graphs and diagrams are subject to very specific 'rules'.

In calculations, always

- quote the equation you are using
- define (or name) any symbols which *you* introduce
- show your working: at least include the substitution of numbers in the equation – this is important, for a mark is often awarded for substitution – as this is a clear indication of whether or not you know what the symbols stand for
- at intermediate stages of a calculation, work to one extra significant figure above the number used for the data in the question.
- at the end of a calculation use the same number of significant figures as are used in the question. Do not confuse decimal places with significant figures. The number of decimal places is not usually important in physics. Whatever you do, do not write down all the digits displayed on your calculator – this will annoy the examiner and you will lose marks.
- give the correct units. The mark for the final answer will either include and require the correct unit, or the units may earn a separate mark
- check your answer. Is it sensible? Could you have made a simple mistake? In particular, check the sign and value of the powers of 10.

In diagrams and graphs, always

- label carefully and fully – especially axes of graphs, where both the variable (what is plotted) and its units are expected
- sketch diagrams, plot points and draw curves on graphs in pencil – you can then rub out any mistakes
- use the space available. Do not plot graphs in a small corner of the graph page. Choose your scales carefully so that the scale is easy to use but the graph occupies most of the page.

General guidance for answering questions

■ Have a clear idea of how much time you should spend on each question before you start to answer it, and monitor the time you take – particularly when doing practice papers.

■ Read the question through completely and carefully before you start to answer it – this way you may avoid setting off on the wrong tack or giving the answer to a later part too early in the question.

■ If the question requires a descriptive answer, consider whether a diagram might help your explanation – even when a diagram is not explicitly asked for.

■ Note the key instruction words in the question. These are: **state**, **explain**, **describe**, **calculate**, **sketch**, etc. Respond to the correct instruction. Do not waste time, for example, on long explanations if the question asks you only to 'state' a law or principle.

The practice papers

Their scope

■ The questions written for these papers cover that part of the syllabus content of all the examination boards which is common and required by the national core syllabus.

■ The questions are set at the full advanced-level standard.

■ Appropriate time has been allocated for the number of questions and the marks awarded.

■ You should use the data and equations booklet or sheet which is provided for the examinations set by your real examination board.

How to work through the practice papers

■ Set aside the correct amount of time and try to ensure that you are not interrupted.

■ Pace yourself – keep a check of how you are using the time.

■ Don't get bogged down on questions you can't do at first sight – leave them and come back to them if you have time.

■ Only look at the answers *after* you have finished.

■ Go back to your text book and/or teacher if you still need more help after seeing the answer.

■ Make a note of the kind of mistakes you made – to help to avoid them in future.

■ Make a list of topics and equations which you need to learn.

How well did you do?

Some guidance about the grade you have achieved for each of the four practice examination papers given in the following table.

Paper	Maximum marks	Grade E = 'pass' mark	Grade C minimum mark	Grade A minimum mark
1	120	45	62	80
2	120	45	62	80
3	67	25	37	50
4	50	20	28	35

Longman
Examination Board

General Certificate of Education

A-level Physics

Paper 1

Time: 3 hours

Instructions

- Attempt ALL questions in this exam paper.
- Answer questions in the spaces provided on this exam paper.
- Spend between 10 and 15 minutes on each question.
- Show all stages in any calculation, and state the units.
- Where diagrams are required, draw and label them clearly.

Information for candidates

- The marks available are shown in brackets after each question or part question.
- This exam paper has 13 questions.
- You are allowed 3 hours for this paper.
- The maximum mark for this paper is 120.

Number	Mark
1.	
2.	
3.	
4.	
5.	
6.	
7.	
8.	
9.	
10.	
11.	
12.	
13.	
Total	

---

1. (a) Give an example of a vector quantity and state the properties which make it a vector.

   ....

   ....

   (2 marks)

   (b) A picture hangs by wire from a nail in the wall, as shown in the diagram.

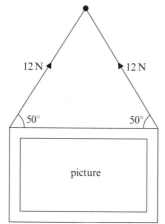

   The tension in both the supporting wires is 12 N.

Leave margin blank

Turn over

1

(i) By resolving forces find the vertical component of the tension in each wire.

....

....

....

(2 marks)

(ii) Calculate the weight of the picture assuming it hangs freely.

....

(1 mark)

(c) If the cross-sectional area of the wire used is $2.0\,\text{mm}^2$, calculate the tensile stress applied to the wire when it supports the picture.

....

....

(2 marks)

[total 7 marks]

2. A bullet of mass $0.024\,\text{kg}$ is fired horizontally from a gun. It leaves the barrel travelling at $250\,\text{m s}^{-1}$.

(a) If the length of the barrel is $0.50\,\text{m}$, calculate:

(i) the average acceleration of the bullet in the barrel,

....

....

....

(2 marks)

(ii) the average force acting on the bullet.

....

....

....

(2 marks)

(b) The barrel of the gun is held $1.6\,\text{m}$ above the ground which is also level. In the following calculations ignore the effects of air resistance.

(i) Calculate the time taken for the bullet to hit the ground after leaving the barrel.

....

....

....

(2 marks)

(ii) Calculate the horizontal range of the bullet.

....

....

(2 marks)

(iii)   Calculate the vertical speed with which the bullet hits the ground.

....

....

....

(2 marks)

(iv)   Calculate the magnitude of the total velocity of the bullet as it strikes the ground.

....

....

....

(1 mark)

[total 11 marks]

3.   The diagram shows a view from above of two air-track vehicles. Vehicle A has a mass of 0.20 kg and vehicle B of 0.12 kg. The vehicles are travelling towards each other at speeds of 2.0 m s^{-1} and 3.0 m s^{-1}. They collide and separate, vehicle A travelling from right to left at 1.2 m s^{-1}.

(a)   (i)   State the principle of conservation of momentum.

....

....

....

(2 marks)

(ii)   What general condition must apply for momentum to be conserved in this collision?

....

....

(1 mark)

(b)   Calculate the velocity of vehicle B after the collision.

....

....

....

(3 marks)

(c)   (i)   State the difference between a *perfectly elastic collision* and an *inelastic collision*.

....

....

(1 mark)

Turn over

(ii) Show that the collision described above is inelastic.

....

....

....

(3 marks)

[total 10 marks]

4. A potter's wheel of diameter 0.30 m spins horizontally about a vertical axis, as shown in the diagram. P is a particle of clay stuck to the edge of the wheel.

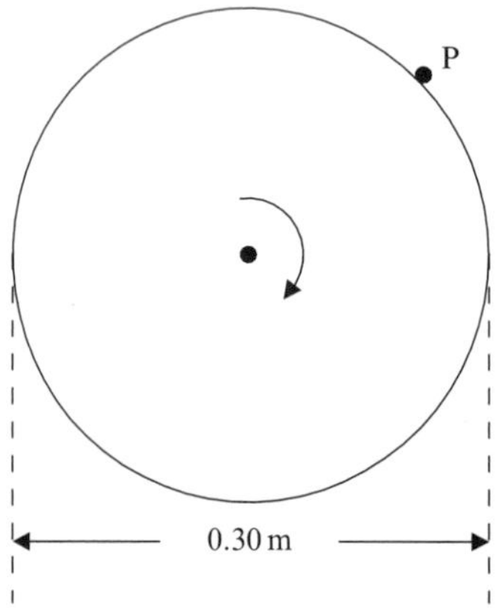

(a) If the wheel rotates at 240 revolutions per minute, calculate

   (i) the angular velocity of the wheel,

....

....

(2 marks)

   (ii) the acceleration of the clay particle P,

....

....

(2 marks)

   (iii) the magnitude of the force acting on P if its mass is 2.0×10^{-3} kg.

....

....

(2 marks)

(b) The maximum radial force at which the clay particle will remain stuck to the side of the wheel is 3.0×10^{-1} N.

   (i) Calculate the angular velocity at which P will leave the wheel.

....

....

....

....

(2 marks)

(ii) At the moment when the clay particle leaves the wheel, what will be its speed and direction?

....

....

....

(3 marks)

[total 11 marks]

5. (a) Define the following terms:

(i) simple harmonic motion,

....

....

....

(3 marks)

(ii) amplitude,

....

....

(1 mark)

(iii) time period.

....

....

(1 mark)

(b) A light helical spring is suspended vertically and has an object of mass 0.050 kg attached to its lower end. The mass is pulled down a small distance and released so that it performs simple harmonic motion with a time period of 2.0 s. Calculate

(i) the spring constant, k,

....

....

....

....

(2 marks)

(ii) the extension of the spring caused by the mass when hanging at rest on the end of the spring.

....

....

(1 mark)

[total 8 marks]

Turn over

6. Graph A below shows the variation of displacement of a particle with distance along the path of a progressive transverse wave at time $t = 0$. The wave is travelling from left to right. Graph B shows the same wave at time $t = 150\,\text{m s}$.

Leave margin blank

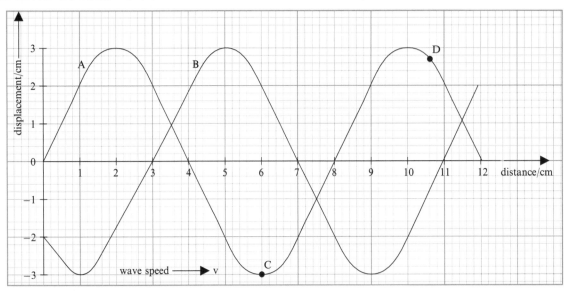

(a) From the graphs, find:

   (i) the wavelength of the progressive wave,

   ....

                                                                       (1 mark)

   (ii) the speed of the wave,

   ....

                                                                       (1 mark)

   (iii) the frequency of the vibrations of a particle in the wave,

   ....

                                                                       (1 mark)

   (iv) the amplitude of the vibrations of a particle in the wave.

   ....

                                                                        (1 mark)

(b) On graph A above mark and label a point where another particle in the wave is:

   (i) in phase with particle D (label, *in phase*),   **(1 mark)**

   (ii) in antiphase with particle D (label, *antiphase*).   **(1 mark)**

(c) On the axes below, sketch a graph of the displacement of the particle labelled C against time from time $t = 0$ to time $t = 1.0\,\text{s}$. Fully label both axes.

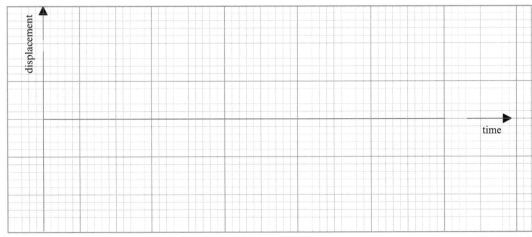

                                                                              (5 marks)

(d) Mark and label points on the graph where the particle C has:

    (i)   maximum speed (label *max. speed*),   **(1 mark)**

    (ii) zero speed (label *zero speed*),   **(1 mark)**

    (iii) maximum acceleration (label *max. acceleration*).   **(1 mark)**

                                                      [total 14 marks]

7. (a) Draw a labelled diagram of the arrangement of the apparatus you would set up in a laboratory to produce and measure double slit interference fringes with monochromatic light.

Give a suitable value for the separation of the two slits: ....

Label with suitable distances any other key distances in your arrangement which would need to be measured in an experiment to find the wavelength of the light.   **(5 marks)**

(b)   (i)   Explain why monochromatic light must be used.

        ....

        ....

        ....

                                                    (1 mark)

  (ii) Describe the appearance of the fringes produced with monochromatic light.

        ....

        ....

        ....

                                                   (2 marks)

  (iii) Describe and explain one change in the observed fringe pattern if the slit separation is reduced.

        ....

        ....

        ....

                                                   (1 mark)

                                                   Turn over

(iv) Describe and explain one change in the observed fringe pattern (other than colour) if the wavelength of the light is reduced.

....

....

....

(1 mark)

(c) Calculate the fringe separation which would be produced by light of wavelength 650 nm using the values you gave for other key dimensions in part (a).

....

....

....

....

(2 marks)

[total 12 marks]

8. (a) A step-index optical fibre consists of a core surrounded by cladding. A light ray strikes the plane end of the core of the fibre at an angle of incidence of 30°, as shown in the enlarged diagram below.

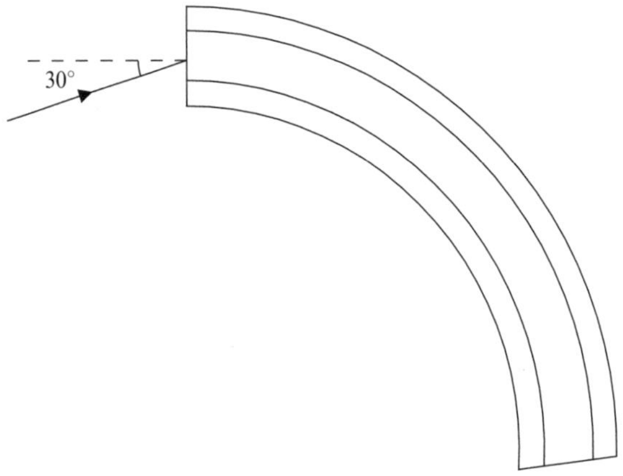

(i) Calculate the angle of refraction of the ray inside the end of the fibre core if the refractive index of the core glass is 1.60.

....

....

....

(2 marks)

(ii) Using a ruler, draw carefully on the diagram an estimated path of the ray as it travels through the fibre. **(2 marks)**

(b) (i) State a suitable value for the refractive index of the cladding.

....

(1 mark)

(ii) Explain why you have suggested this value.

....

....

(2 marks)

[total 7 marks]

9. In the circuits below the bulbs are identical and the batteries have negligible internal resistance.

Circuit A

Circuit B

(a) For circuit A:

   (i) calculate the current in each bulb,

....

....

(1 mark)

   (ii) calculate the power of each bulb.

....

....

(1 mark)

(b) For circuit B:

   (i) calculate the current in each bulb,

....

....

(2 marks)

   (ii) calculate the p.d. across each bulb.

....

....

(1 mark)

(c) Describe the brightness of the bulbs if they are all rated at 0.8 A and 2.5 V

   (i) in circuit A, ...

....

(1 mark)

Turn over

Leave margin blank

(ii) in circuit B. ....

....

(1 mark)

(d) If the filament of one of the bulbs in circuit B breaks, describe and explain what happens to the brightness of the other two bulbs.

....

....

....

(2 marks)

[total 9 marks]

10. The circuit below is connected up with switch S closed. The variable resistor, R is adjusted so that the microammeter reads $100\,\mu A$. For this question you should assume that the micro-ammeter has negligible resistance.

(a) With switch S closed,

(i) calculate the resistance of the circuit connected to the battery,

....

....

(1 mark)

(ii) state the p.d. across the capacitor. ....

(1 mark)

(b) When switch S is opened, but the circuit is not adjusted, state briefly what happens.

....

....

....

(2 marks)

(c) After switch S is opened the variable resistor, R is gradually adjusted to maintain a constant current of $100\,\mu A$. This current is sustained for $10\,s$. After $10\,s$ no more current flows.

(i) Calculate the charge stored in the capacitor at the moment the switch was opened.

....

....

(1 mark)

(ii)   Calculate the capacitance of the capacitor.

....

....

(1 mark)

[total 6 marks]

11. The circuit below shows a battery of e.m.f. 12 V and internal resistance $2.0\,\Omega$ connected to two resitors R_a and R_b with the resistances shown.

Calculate

(i)   the current in the circuit,

....

....

(1 mark)

(ii)   the readings on the voltmeters,

V_1 ....

V_2 ....

(2 marks)

(iii)   the terminal p.d. of the battery when connected to this circuit.

....

(1 mark)

(iv)   Explain your answer to part (iii).

....

....

(2 marks)

[total 6 marks]

12. A $10\,\mu$F capacitor is fully charged by connecting it to a 24 V battery.

(a)   (i)   How much charge would be stored in the capacitor?

....

....

(1 mark)

Turn over

    (ii)   How much energy would be stored in the capacitor?

....

....

(1 mark)

    (iii)   How much energy would be required from the battery to charge the capacitor to the 24 V level?

....

(1 mark)

    (iv)   Explain why your answer to part (iii) is different from part (ii).

....

....

....

(2 marks)

(b)   The fully charged capacitor in (a) is now disconnected from the battery and connected across another identical capacitor.

    (i)   How much charge will now be stored on each capacitor?

....

(1 mark)

    (ii)   What will be the total energy stored in the two capacitors?

....

....

....

(1 mark)

    (iii)   Explain why your answer to (b) (ii) is different from your answer to (a) (ii).

....

....

....

(2 marks)

[total 9 marks]

13. A radioactive source emits a beam of alpha, beta and gamma radiation.

(a)   Explain how you would remove alpha radiation from the beam without affecting the beta or gamma radiation very much.

....

....

(1 mark)

(b) A magnetic field is then applied to the beam to separate the beta and gamma radiation. Sketch a diagram to show the paths of the beta and gamma radiation before, during and after entering the region of the magnetic field. Indicate clearly the direction of the magnetic field and both the directions and shapes of the paths of the beta and gamma radiations.

(5 marks)

(c) (i) What would be the effect on the paths of the beta and gamma radiations of increasing the strength of the magnetic field?

....

....

(1 mark)

(ii) Why would the magnetic field used to separate the beta and gamma radiations from the beam not be effective in also separating the alpha radiation from the other two?

....

....

(1 mark)

(d) Calculate the magnitude of the force acting on an electron travelling at $2.0 \times 10^6 \, \text{m s}^{-1}$ through a magnetic field of magnetic flux density $4.0 \times 10^{-4} \, \text{T}$ at $90°$ to the field direction. The charge of the electron is $1.6 \times 10^{-19} \, \text{C}$.

....

....

....

(2 marks)

[total 10 marks]

Total: 120 marks

Longman Examination Board

General Certificate of Education

A-level Physics

Paper 2

Time: 3 hours

Instructions

- Attempt ALL questions in this exam paper.
- Answer questions in the spaces provided on this exam paper.
- Spend between 10 and 15 minutes on each question.
- Show all stages in any calculation, and state the units.
- Where diagrams are required, draw and label them carefully.

Information for candidates

- The marks available are shown in brackets after each question or part question.
- This exam paper has 14 questions.
- You are allowed 3 hours for this paper
- The maximum mark for this paper is 120.

Number	Mark
1.	
2.	
3.	
4.	
5.	
6.	
7.	
8.	
9.	
10.	
11.	
12.	
13.	
14.	
Total	

---

1. (a) The diagram below shows a cross-section of a solenoid which carries a current in the direction indicated. Draw on the diagram the associated magnetic field pattern inside and around the solenoid, indicating its direction.

Leave margin blank

⊗⊗⊗⊗⊗⊗⊗⊗⊗⊗⊗⊗⊗⊗⊗⊗⊗⊗⊗⊗

⊙⊙⊙⊙⊙⊙⊙⊙⊙⊙⊙⊙⊙⊙⊙⊙⊙⊙⊙⊙

⊗ = current down
⊙ = current up

(4 marks)

   (b) (i) On the axes below sketch a curve to show how the extension of an elastic material varies with increasing stretching load. Label the elastic part of the curve *elastic*. Also label key features of the curve. **(2 marks)**

(ii) Extend the curve to show how the extension becomes plastic with increasing load. Label this part of the curve *plastic*. Also label key features of this part of the curve.

(2 marks)

extension

load

[total 8 marks]

2. (a) An electric motor raises a mass of 5 kg a vertical height of 30 m in 40 s at a constant speed. If it is 25% efficient in converting the electrical power input into mechanical power output, calculate the required input electrical power.

....

....

....

....

....

(3 marks)

(b) What happens to the other 75% of the input electrical power?

....

....

(1 mark)

[total 4 marks]

3. (a) What length of eureka wire of diameter 0.10 mm is required to make a coil of 20 ohm resistance? The resistivity of eureka alloy is $4.9 \times 10^{-7}\,\Omega\mathrm{m}$.

....

....

....

....

....

(3 marks)

Turn over

(b) A uniform tube containing a mercury column passes a current of 0.1 A from top to bottom through the column when it is connected to a source of p.d. of low internal resistance. If all of this mercury is now put into another uniform tube which has twice the radius of the first tube, what current will flow when the same p.d. is applied?

....

....

....

....

....

(3 marks)

[total 6 marks]

4. A radio transmitter emits a radio wave at a power of 400 kW on a wavelength of 400 m.
The Planck constant $= 6.6 \times 10^{-34}$ J s.
The speed of radio waves $= 3.0 \times 10^8$ m s^{-1}.
Calculate

(a) the frequency corresponding to this wavelength,

....

....

(2 marks)

(b) the energy of a radio photon of this frequency in joules,

....

....

(2 marks)

(c) the number of photons emitted per second by the transmitter.

....

....

(2 marks)

[total 6 marks]

5. (a) The gravitational field produced by a planet is radial, as shown in the figure below. On the axes alongside, sketch a curve to show how the gravitational potential above the surface of the planet V varies with the distance r from the centre of the planet. The planet has a radius R.

(3 marks)

(b) The electric field between the charged plates of a parallel-plate capacitor is described as being uniform at its centre.

   (i) On the diagram below, sketch the shape of the electric field between the plates both at its centre and near the edges of the plates, indicating any key features and the direction of the field. **(3 marks)**

   (ii) On the axes below, show how the electric potential, V varies with distance, y from the lower plate. **(2 marks)**

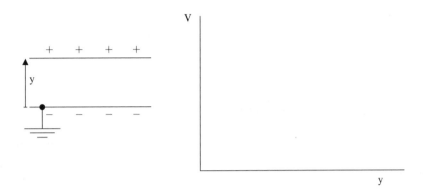

[total 8 marks]

6. (a) (i) Explain what is meant by the *root mean square speed* of the molecules in a gas.

   ....

   ....

   ....

   (1 mark)

   (ii) State the factor or factors on which the mean kinetic energy of a gas molecule depends.

   ....

   ....

   (2 marks)

(b) Four molecules, A, B, C and D, each of mass 4.2×10^{-26} kg, move with the speeds and in the directions shown in the figure below.

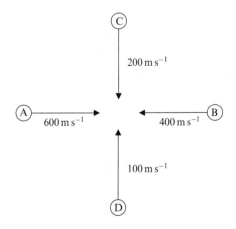

Turn over

Calculate for these molecules,

    (i)   the mean momentum of molecules A and B,

....

....

....

(1 mark)

    (ii)   the mean momentum of molecules C and D,

....

....

....

(1 mark)

    (iii)   the mean momentum of all four molecules,

....

....

....

(3 marks)

    (iv)   the mean kinetic energy of all four molecules.

....

....

....

....

....

(4 marks)

[total 12 marks]

7.   **(a)**   In an experiment with a vacuum photoelectric cell, the cell was connected in series with a microammeter. When the metal cathode of the cell was illuminated with monochromatic red light of wavelength 660 nm, no current was recorded by the microammeter. However, when the light was changed to monochromatic blue light of the same intensity and of wavelength 420 nm, a steady small current was registered on the microammeter.

Explain these observations using the photon theory of light.

....

....

....

....

....

....

....

....

(3 marks)

(b) (i) Calculate the photon energy of the blue light in joules.
The speed of light $= 3.0 \times 10^8 \, \text{m s}^{-1}$.
The Planck constant $= 6.6 \times 10^{-34} \, \text{J s}$.

....

....

....

....

....

(2 marks)

(ii) Calculate the maximum kinetic energy of an electron released from the metal surface by the blue light if the work function of the metal surface is 2.40 eV.

....

....

....

....

....

(3 marks)

[total 8 marks]

8. (a) A stream of electrons is accelerated by an electron gun in an evacuated tube. If the p.d. between anode and cathode is 4.0 kV, calculate the speed of the electrons as they leave the gun. The charge of the electron is $1.6 \times 10^{-19} \, \text{C}$ and its mass is $9.1 \times 10^{-31} \, \text{kg}$.

....

....

....

....

(3 marks)

(b) The stream of electrons in part (a) then enters the electric field between two parallel plates, as shown in the figure below.

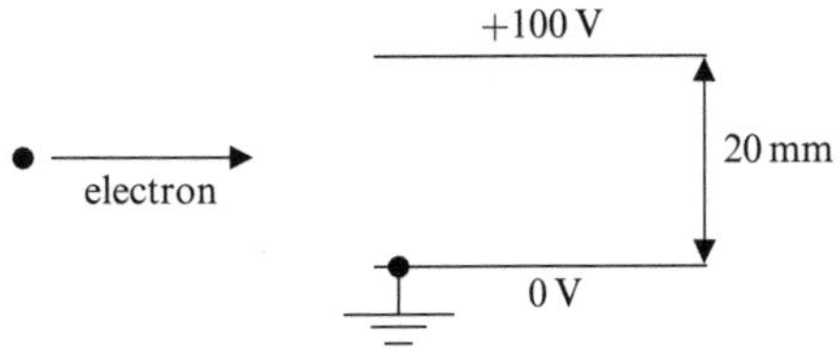

(i) Sketch carefully on the diagram the path of the electrons as they pass through and beyond the electric field. **(2 marks)**

(ii) Calculate the electric field strength between the plates.

....

....

(2 marks)

Turn over

(iii) Calculate the magnitude of the electrostatic force on each electron while in the field.

....

....

(2 marks)

(c) The stream of electrons in (a) on another occasion entered the region of the uniform magnetic field shown in the figure below.

region of magnetic field perpendicular to plane of paper and downwards
$B = 4.0 \times 10^{-2}$ T

electrons

(i) Sketch carefully on the diagram the path of the stream of electrons as it passes through and beyond the region of the magnetic field. **(2 marks)**

(ii) Calculate the magnitude of the force acting on each electron in the stream.

....

....

(2 marks)

(d) (i) Give one significant difference between the shapes of the paths of the electron beams while in the two different kinds of fields.

....

....

....

(2 marks)

(ii) State what would happen to the path of the electron beam in the magnetic field if the field was set at 80° to the path of the electron beam, rather than 90°.

....

....

(1 mark)

[total 14 marks]

9. (a) State Lenz's law of electromagnetic induction.

....

....

....

(2 marks)

(b) With the aid of a diagram, describe a simple experiment which you could use to demonstrate Lenz's law. Explain how the observations made in the experiment are consistent with Lenz's law.

....

....

....

....

....

....

....

....

(6 marks)

[total 8 marks]

10. (a) (i) Explain what is meant by a *thermal neutron* in a nuclear reactor.

....

(1 mark)

(ii) Explain why a moderator is used in a thermal nuclear reactor.

....

....

(1 mark)

(iii) Suggest a suitable material for use as a moderator and state one essential property it must have to work well as a moderator.

....

....

(2 marks)

(iv) Explain the function of control rods in a thermal nuclear reactor.

....

....

(1 mark)

Turn over

(v) Suggest a suitable material for use as a control rod and state one essential property it must have to perform the basic function of a control rod.

....

....

(2 marks)

(b) Describe the process by which heat is generated in the core of a thermal nuclear reactor.

....

....

....

....

....

....

....

....

....

(5 marks)

[total 12 marks]

11. Monochromatic light of wavelength 6.0×10^{-7} m is incident at right-angles on the surface of a diffraction grating.

(a) Calculate the spacing of the lines on the diffraction grating if the first order maximum is observed at an angle of 24°.

....

....

....

(2 marks)

(b) Calculate the angle at which you would expect to observe the first order and second order maxima for blue light of wavelength 4.8×10^{-7} m.

....

....

....

(2 marks)

(c) What is the total number of orders or bright maxima of the blue light which you would expect to see in the light transmitted by this grating?

....

....

(2 marks)

[total 6 marks]

12. (a) The mass of a particular radioisotope in a sample is initially 6.4×10^{-3} kg. After 42 days what is left of this isotope is separated from the sample and found to have a mass of 1.0×10^{-4} kg.

Calculate

   (i) the half-life of the isotope,

....

....

....

....

(2 marks)

   (ii) the decay constant of the isotope.

....

....

....

(2 marks)

(b) When waste radioactive material from a nuclear reactor is being processed, waste isotopes with long half-lives must be stored safely. Describe briefly some of the safety procedures and techniques which are used to store these long half-life materials safely.

....

....

....

....

....

....

....

(4 marks)

[total 8 marks]

13. (a) With reference to waves on a string, state two differences between travelling waves and stationary waves.

   1. ....

....

   2. ....

....

(2 marks)

(b) The figure below shows a string vibrating in its fundamental mode.

Turn over

(i)   What is meant by the statement that this is the *fundamental mode* of vibration?

....

....

(1 mark)

(ii)   By making measurements on the diagram, find the wavelength and maximum amplitude of the standing wave. Assume that the diagram is drawn life-size.

Wavelength = ...

Maximum amplitude = ...

(2 marks)

(c)   On the diagram, draw the form of the standing wave which would have twice the frequency of the fundamental mode of vibration. **(1 mark)**

[total: 6 marks]

14. (a)   The diagram below shows an incomplete electromagnetic spectrum. Complete the spectrum by placing in their ascending wavelength order the other recognised regions of the spectrum.

(4 marks)

(b)   Below each section of the spectrum which you have labelled state a typical wavelength for radiation in that part of the spectrum. A value giving the order of magnitude in metres is adequate. **(4 marks)**

(c)   (i)   State two properties which are common to all the regions of the electromagnetic spectrum.

1. ....

2. ....

(2 marks)

(ii)   State two features, other than wavelength and frequency, of the regions of the spectrum which change as we pass along the spectrum from left to right.

1. ....

2. ....

(2 marks)

[total 12 marks]

Total: 120 marks

Longman Examination Board

General Certificate of Education

A-level Physics

Paper 3 (Graphical questions)

Time: 2 hours

Number	Mark
I.	
2.	
3.	
Total	

Instructions

- Attempt ALL questions.

- Answer questions in the spaces provided on this exam paper.

- Spend up to 40 minutes on each question.

- Show all stages in any calculation, and state the units.

- Where diagrams are required, draw and label them clearly.

Information for candidates

- The marks availabe are shown in brackets after each question or part question.

- This exam paper has 3 questions.

- You are allowed 2 hours for this paper.

- The maximum mark for this paper is 67.

Leave margin blank

1. The table below gives the results of an experiment in which a steel cable is stretched until it breaks. The original length of the cable was 30.0 mm and its initial cross-sectional area was $1.0 \times 10^{-5}\,\text{m}^2$. The cable is subjected to a controlled stretching force which is gradually adjusted so that the cable continues to stretch steadily. The data shows how the stretching force varied as the cable was stretched.

extension/mm	0	0.05	0.10	0.15	0.20	0.25	0.30	0.35	0.40	0.45
force/kN	0	3.2	6.4	9.8	13.0	12.0	13.6	14.4	15.2	16.0

extension/mm	0.50	0.55	0.60	0.65	0.70	0.75	0.80	0.85	0.90	0.95
force/kN	16.6	17.2	17.6	18.0	18.0	17.8	17.6	17.0	16.0	14.0

Turn over

25

(a) On the graph page opposite plot a graph of force (on the *y*-axis) against extension (on the *x*-axis). Draw the best fit curve through the plotted points. **(8 marks)**

(b) (i) State Hooke's law.

....

....

....

(2 marks)

(ii) Mark on the graph you have plotted the part of the curve for which the specimen obeyed Hooke's law and label the limit of proportionality with an L.

(2 marks)

(c) (i) If the process of stretching had been stopped when the extension had reached 0.20 mm and the stretching force removed, what would have been the total length of the cable after the stretching? Give an explanation for your answer.

....

....

....

(2 marks)

(ii) If the process of stretching had been continued only until the extension had reached 0.50 mm, draw on your graph what you would expect to have happened to the extension as the force was reduced to zero. Label any important features of the curve you have drawn. **(3 marks)**

(d) In the region of the curve where Hooke's law is followed, calculate the Young Modulus for the specimen.

....

....

....

....

....

....

(6 marks)

(e) Calculate the energy stored in the cable when the extension is 0.10 mm.

....

....

....

(3 marks)

(f)   In terms of the structure of the steel, explain what happens

    (i)   while the stretching obeys Hooke's law,

        ....

        ....

        ....

        (2 marks)

    (ii)   when the curve extends beyond the limit of proportionality.

        ....

        ....

        ....

        (2 marks)

[total 30 marks]

Turn over

2. (a) A diffraction grating is mounted on the turntable of a spectrometer so that its surface is normal to the light reaching it from the collimator. In setting up the spectrometer to view the line spectrum of a light source, what adjustments should be made to

   (i) the collimator,

....

....

(2 marks)

   (ii) the telescope?

....

....

(2 marks)

(b) The light source emits four discrete wavelengths which are listed in the table below. The table also gives the readings obtained from the spectrometer when its telescope cross-wires were lined up accurately on each of the first-order diffraction maxima. Readings were taken with the telescope set both to the left and the right of the central diffraction maximum which was observed at 180.0°.

Wavelength/nm	Telescope left°	Telescope right°		
448	165.7	194.4		
501	164.0	196.0		
588	161.1	198.9		
668	158.4	201.6		

Use the data in the table to draw a straight-line graph from which you can find the number of lines per metre on the diffraction grating.

....

....

(8 marks)

(c) The light source was replaced by a monochromatic one. Readings were obtained for the **second**-order diffraction maxima using the same grating.

Telescope left°	Telescope right°
136.8	223.2

Calculate

(i) the wavelength of the light emitted by the monochromatic source,

....

....

....

(2 marks)

(ii) the total number of diffraction maxima visible on each side of the central maximum.

....

....

....

(2 marks)

[total 16 marks]

3. A source of gamma radiation of long half-life was mounted with its open side facing a Geiger–Müller tube, as shown in the figure below. The distance, d, between the front of the source housing and the side wall of the Geiger–Müller tube could be varied.

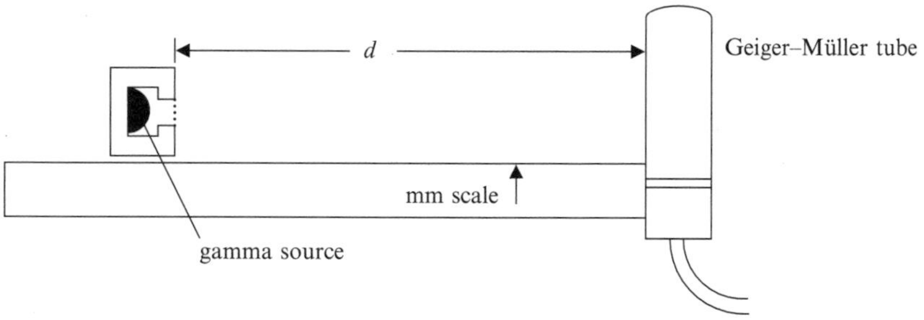

(a) Before the gamma source was installed several background count-rate measurements were made, each over a 10-minute period.

Counts in 10 minutes	580	570	526	536	546	566	554	534	490	498

Calculate the mean background count-rate in counts per second.

....

....

(2 marks)

Turn over

(b) The gamma source was installed and the count, *N*, recorded, over a 10-minute period, for each of various distances, *d*. The results are recorded in the table below.

d/mm	15	30	45	60	75	90	105
N (10 minutes)	1 139 404	495 010	265 802	226 793	115 198	84 608	65 401
Count-rate per second							
Corrected count-rate per second, *c*							
$\dfrac{1}{\sqrt{c}}$							

(i)   Complete the data in the table.

(The corrected count-rate is corrected for the mean background level of radiation.)

(4 marks)

(ii) Fill in the missing units in the lowest left-hand cell of the table.   **(1 mark)**

(c) Plot the graph of $\dfrac{1}{\sqrt{c}}$ (on the *y*-axis) against *d* (on the *x*-axis) on the graph paper below.

(6 marks)

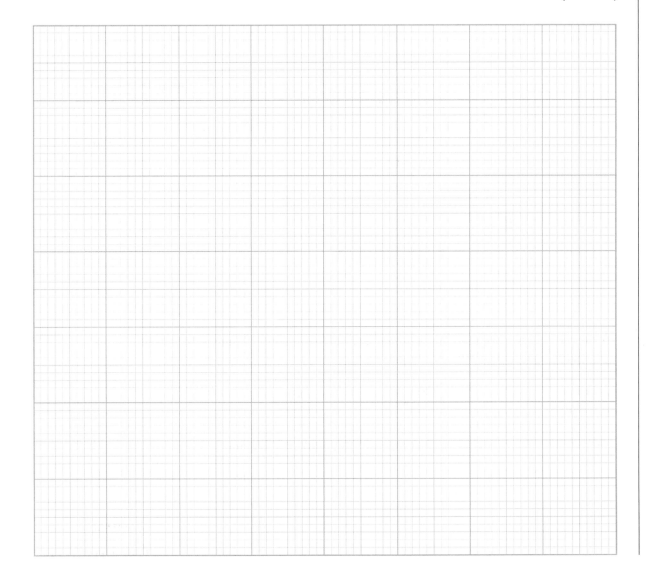

(d) (i) Find the gradient of the graph.

....

....

....

(3 marks)

(ii) What can you conclude from the fact that this plot gives a straight-line graph of the gradient you have calculated? Explain your conclusions.

....

....

....

....

....

(3 marks)

(iii) There is an intercept on the *x*-axis. What is its value? ....

What does this intercept tell you? ....

....

....

(2 marks)

[total 21 marks]

Total: 67 marks

Longman Examination Board

General Certificate of Education

A-level Physics

Paper 4 (Experimental questions)

Time: 2 hours

Number	Mark
1.	
2.	
3.	
Total	

Instructions

- Attempt ALL questions.

- Answer questions in the spaces provided on this exam paper.

- Spend up to 40 minutes on each question.

- Show all stages in any calculation, and state the units.

- Where diagrams are required, draw and label them clearly.

Information for candidates

- The marks availabe are shown in brackets after each question or part question.

- This exam paper has 3 questions.

- You are allowed 2 hours for this paper.

- The maximum mark for this paper is 50.

Leave margin blank

1. The core of a 'lead' pencil is made of a non-metallic material which has, at 25° C, a resistivity of 4.20×10^{-3} Ωm.

    (a) If the pencil core is 120 mm long and its mean diameter is 1.50 mm, calculate the electrical resistance of the pencil between its ends.

    ....

    ....

    ....

    ....

    (4 marks)

    (b) You are provided with a specimen of the same pencil 'lead' which is about 40 mm long and completely stripped of the surrounding wood. Describe experiments you would carry out to confirm that the resistivity of the material has the value given in part (a). You are not allowed to use a digital multimeter capable of measuring resistance in ohms directly.

(i) State the measurements you would make and list the instruments you would you use.

....

....

....

....

(4 marks)

(ii) Draw a labelled circuit diagram of the circuit you would use.   **(2 marks)**

(iii) Give an account of how you would process the measurements made to calculate the resistivity of the material.

....

....

....

....

....

(3 marks)

(c) The pencil 'lead' material has a negative temperature coefficient of resistance.

(i) What is meant by the statement that the coefficient is 'negative'?

....

(1 mark)

(ii) Give one possible explanation of why this material behaves in this way.

....

....

....

(2 marks)

Turn over

(d)   A different specimen of pencil 'lead' of resistance 100 Ω was used in a similar experiment, and a steady current of 0.5 A was passed through it.

   (i)   Calculate the rate of conversion of electrical energy into heat.

   ....

   ....

   ....

   (2 marks)

   (ii)   Comment on whether you think the piece of 'lead' would feel very hot to touch.

   ....

   ....

   ....

   (2 marks)

   [total 20 marks]

2.   (a)   For each of the following terms explain its meaning and give an example of it occurring in a vibrating mechanical system or object.

   (i)   Natural vibration ....

   ....

   ....

   (2 marks)

   (ii)   Forced vibration ....

   ....

   ....

   (2 marks)

   (iii)   Resonance ....

   ....

   ....

   (3 marks)

   (b)   Describe a demonstration which you could set up in a laboratory with a mechanical system which would illustrate the meaning of the three terms above.

   (i)   *Labelled diagram of the apparatus.*   **(4 marks)**

(ii) *Explain how the apparatus is used and what you would expect to see happen.*

....

....

....

....

(3 marks)

[total 14 marks]

3. (a) Define the term 'efficiency' as it would apply to an electric motor.

....

....

....

(2 marks)

(b) Describe an experiment you would perform in a laboratory to measure the efficiency of an electric motor as it is used to raise a load at a steady rate. You can imagine that the experiment is a small-scale model of an electrically operated lift.

(i) Draw a labelled diagram of the apparatus as you would assemble it.

(3 marks)

(ii) Draw an electric circuit diagram to show how the power supply to the motor would be controlled and measured. **(2 marks)**

Turn over

(iii) List all the measurements which you would make and specify the apparatus you would use to make each one.

....

....

....

....

....

(5 marks)

(iv) Explain how the measurements would be used to calculate the efficiency of the motor.

....

....

....

....

....

....

(2 marks)

(c) The efficiency of a particular small electric motor was found to be only 20% when lifting a small load, but increased to 25% when lifting a larger load.

(i) Give a possible reason why the efficiency of the small motor was found to be so low.

....

....

....

(1 mark)

(ii) Give a reason why the efficiency increased when the load was increased.

....

....

....

(1 mark)

[total 16 marks]

Total: 50 marks

Solutions to practice exam papers

Each * represents 1 mark.

When you have finished marking your answers, refer to the front of the book for guidance on the grade you have achieved in each practice exam paper.

Solutions to Paper 1

1. (a) Examples include: force, displacement, velocity, acceleration, momentum *

    A vector has both magnitude (size) and direction * **2 marks**

    (b) (i) Vertical component in each wire $= 12 \cos 40$ * $= 9.2\,\text{N}$ * **2 marks**

    (ii) Weight of picture $= 2 \times 9.2 = 18.4\,\text{N}$ * **1 mark**

> **TIP**
> Note that in all numerical answers, the final figure must include correct units. However, two marks would not be lost for omitting the same unit from two answers in the same question.

    (c) Tensile stress $= \dfrac{force}{area} = \dfrac{12}{2.0 \times 10^{-6}}$ * $= 6.0 \times 10^{6}\,\text{N}\,\text{m}^{-2}$ * **2 marks**

> **TIP**
> The first mark is for the correct substitution, but the key figure to get correct is the area in m^2 units; remember that there are $10^3\,\text{mm}$ in 1 m but $10^3 \times 10^3 = 10^6\,\text{mm}^2$ in $1\,\text{m}^2$.

[total 7 marks]

2. (a) (i) $a = \dfrac{v^2}{2s} = \dfrac{250^2}{2 \times 0.50}$ * $= 250^2 = 62\,500$ or $6.25 \times 10^4\,\text{m}\,\text{s}^{-2}$ * **2 marks**

> **TIP**
> The equation is obtained by rearranging $v^2 = u^2 + 2as$ in which the initial velocity, u, is zero.

    (ii) $F = ma = 0.024 \times 6.25 \times 10^4$ * $= 1.5 \times 10^3\,\text{N}$ * **2 marks**

    (b) (i) $t = \sqrt{\dfrac{2s}{a}} = \sqrt{\dfrac{2 \times 16}{9.8}}$ * $= 0.57\,\text{s}$ * **2 marks**

> **TIP**
> The equation is obtained from $s = ut + \frac{1}{2}at^2$. Put $u = 0$ and rearrange.
> Note that the vertical motion is independent of the horizontal motion. The bullet still falls freely with gravity with a vertical acceleration of $9.8\,\text{m}\,\text{s}^{-2}$, regardless of its horizontal speed.

    (ii) For the horizontal motion:

    $s = vt = 250 \times 0.57$ * $= 14\,\text{m}$ * (2 significant figures only required) **2 marks**

    (iii) $v = u + at = 0 + 9.8 \times 0.57$ * $= 5.6\,\text{m}\,\text{s}^{-1}$ * **2 marks**

    (iv) Using vector addition at right-angles:

    resultant velocity^2 = vertical velocity^2 + horizontal velocity^2

    resultant velocity$^2 = 5.6^2 + 250^2$

    resultant velocity $= 250\,\text{m}\,\text{s}^{-1}$. * (To two significant figures the small vertical velocity component makes no difference to the much larger horizontal component.)

1 mark

[total 11 marks]

3. (a) (i) In all collisions in which no external forces act *, the total momentum before the collision equals the total momentum after the collision. *
2 marks

(ii) The air track must provide extremely low friction conditions for the external forces to be negligible. *
1 mark

(b) Total momentum before the collison = momentum of vehicle A + momentum of vehicle B

> Remember that momentum is a vector quantity. It is vital to take the direction of motion into account. It is usual to take velocities to the right as positive and those to the left as negative. Momentum is mass × velocity.

Total momentum before the collision = $(0.20 \times 2.0) + (0.12 \times -3.0)$

[B moving to the left]

$= 0.4 - 0.36 = 0.04 \, \text{kg m s}^{-1}$ *

Total momentum after the collision = $(0.20 \times -1.2) + (0.12 \times v)$ * $= 0.04$

$= -0.24 + 0.12v = 0.04$

Rearranging gives $v = 2.3 \, \text{m s}^{-1}$. Vehicle B moves to the right. *
3 marks

(c) (i) In a perfectly elastic collision all kinetic energy is conserved. In an inelastic collision some kinetic energy is converted into other forms of energy such as heat and sound. *
1 mark

(ii) Kinetic energy before collision = $\frac{1}{2}mv^2$ for vehicle A $+ \frac{1}{2}mv^2$ for vehicle B

$= (\frac{1}{2} \times 0.20 \times 2.0^2) + (\frac{1}{2} \times 0.12 \times 3.0^2)$

$= 0.40 + 0.54 = 0.94 \, \text{J}$ *

> Note that kinetic energy is a scalar quantity and is always positive, so there are no minus signs to worry about when doing a kinetic energy calculation.

Kinetic energy after collision = $\frac{1}{2}mv^2$ for vehicle A $+ \frac{1}{2}mv^2$ for vehicle B

$= (\frac{1}{2} \times 0.20 \times 1.2^2) \times (\frac{1}{2} \times 0.12 \times 2.3^2)$

$= 0.144 + 0.317 = 0.46 \, \text{J}$ *

The total kinetic energy has fallen from 0.94 J to 0.46 J therefore the collision is inelastic. *
3 marks

[total 10 marks]

4. (a) (i) Angular velocity, $\omega = \dfrac{angle}{time} = \dfrac{240 \times 2\pi}{60}$ * $= 25 \, \text{rad s}^{-1}$ *
2 marks

(ii) Acceleration, $a = r\omega^2 = 0.15 \times 25^2$ * $= 94 \, \text{m s}^{-2}$ *
2 marks

(iii) Force $= ma = 2.0 \times 10^{-3} \times 94$ * $= 1.9 \times 10^{-1}$ or $0.19 \, \text{N}$ *
2 marks

(b) (i) Using $F = mr\omega^2$

$3.0 \times 10^{-1} = 2.0 \times 10^{-3} \times 0.15 \times \omega^2$ * (where ω is the maximum angular velocity)

Gives $\omega = 32 \, \text{rad s}^{-1}$ *
2 marks

(ii) Speed $= r\omega = 0.15 \times 32$ * $= 4.8 \, \text{m s}^{-1}$ * at a tangent to the rim of the wheel. *
3 marks

[total 11 marks]

5. (a) (i) A particle or oscillator has simple harmonic motion if it oscillates symmetrically about a fixed point and has the same period for all amplitudes. * For all displacements from the mid point it has an acceleration towards that point * which is directly proportional to its displacement. *   **3 marks**

   (ii) Amplitude is the maximum displacement from the mid point of the oscillation (or from the equilibrium position). *   **1 mark**

   (iii) Time period is the time taken for one complete cycle of the oscillation (or return journey from and to either extreme position). *   **1 mark**

(b) (i) $T = 2\pi\sqrt{\dfrac{m}{k}}$ rearranged gives:

$$k = \frac{4\pi^2 m}{T^2} = \frac{4\pi^2 \times 0.050}{2.0^2} * = 0.49\,\text{m N}^{-1} *$$ **2 marks**

   (ii) Extension, $\Delta l = kF = 0.49 \times 0.050 \times 9.8 = 0.24\,\text{m}$ *   **1 mark**

                                                                                                   [total 8 marks]

6. (a) (i) wavelength $= 8.0\,\text{cm}$ * (A crest and a trough)   **1 mark**

   (ii) speed $= \dfrac{s}{t} = \dfrac{3 \times 10^{-2}\,\text{m}}{150 \times 10^{-3}\,\text{s}} = 0.2\,\text{m s}^{-1}$ *   **1 mark**

   (iii) $T = \frac{8}{3} \times 150 \times 10^{-3}\,\text{s} = 0.40\,\text{s}$

   (The time for the wave to travel a whole wavelength of 8 cm.)

   $f = \dfrac{1}{T} = \dfrac{1}{0.40} = 2.5\,\text{Hz}$ *   **1 mark**

   (iv) amplitude $= 3.0\,\text{cm}$ *

   (The maximum displacement from the equilibrium position, not the peak-to-peak value.)   **1 mark**

(b)

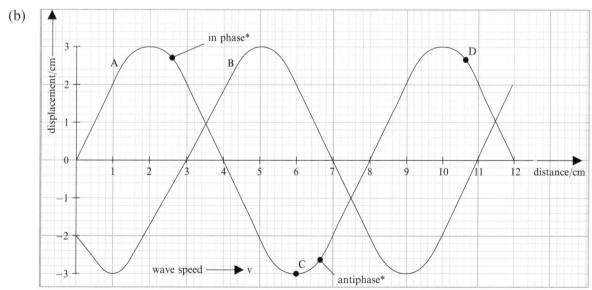

   (i) * for correct in phase   **1 mark**

   (ii) * for correct antiphase   **1 mark**

(c)

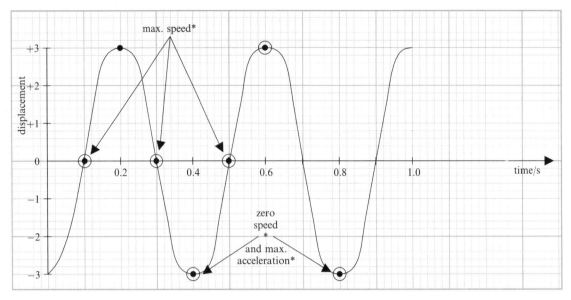

* for correct scales on both axes
* for the general sinusoidal shape
* for showing 2.5 complete oscillations in the 1.0 seconds
* for the correct amplitude of 3.0 cm
* for the curve starting at the correct displacement of −3 cm   **5 marks**

(d)   *** for the three correct labels as shown on the diagram above.   **3 marks**

[total 14 marks]

7. (a)

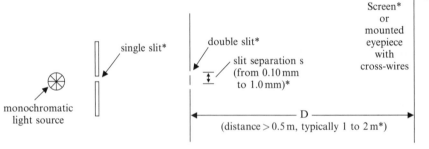

* for single slit
* for double slit
* for screen
* for a value for D, the distance between the double slits and the screen or eyepiece, from about 0.5 m to about 2.0 m
* for a value of s, the slit separation, from 0.10 mm to 1.0 mm   **5 marks**

(b)   (i)   Fringe separation is directly proportional to wavelength, therefore fringe patterns overlap for light with different wavelengths and become confused – so measurements are not possible. *   **1 mark**

(ii) Fringes have equal spacing *

Fringes are not sharp – their intensity (brightness) varies smoothly between dark fringes and bright fringes. *   **2 marks**

TIP

This is an example of a written answer where it might be quite hard to get both marks if you do not explain yourself clearly. An alternative to this would be a simple labelled sketch diagram which would certainly earn the marks and is actually a more detailed and knowledgeable answer.

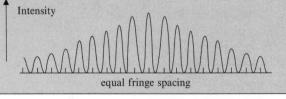

equal fringe spacing

(iii) Fringe spacing increases because fringe spacing is inversely proportional to the slit separation. *

(This is why slits very close together are used – to give a good spreading out of the fringes.) **1 mark**

(iv) Fringe spacing reduces (fringes closer together) – because the spacing is directly proportional to the wavelength of the light. * **1 mark**

(c) Fringe spacing $= \dfrac{\lambda D}{s} = \dfrac{650 \times 10^{-9} \times 1.0}{5.0 \times 10^{-4}}$ * (using $D = 1.0\,\text{m}$ and $s = 0.50\,\text{mm}$)

$= 1.3 \times 10^{-3}\,\text{m}$ or $1.3\,\text{mm}$ * **2 marks**

[total 12 marks]

8. (a) (i) The formula for refractive index, $n = \dfrac{\sin i}{\sin r}$, rearranged gives:

$\sin r = \dfrac{\sin i}{n} = \dfrac{\sin 30}{1.60} = \dfrac{0.50}{1.60}$ * $= 0.3125$

so angle $r = 18.2°$ * **2 marks**

(ii)

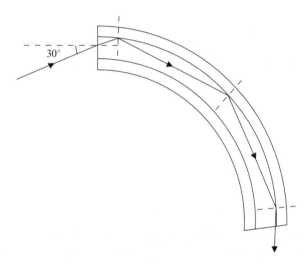

* for internal reflections occurring round the internal outside edge of the core
* for angles of incidence and reflection looking equal. **2 marks**

(b) (i) Any value between 1.5 and 1.2 * **1 mark**

(ii) For **total internal reflection** to occur at an **inside surface** * (boundary) between two types of glass the outer material must have a significantly **lower refractive index**. *

(The words in bold, or very similar words, are key to earning the two marks.) **2 marks**

[total 7 marks]

9. (a) (i) $I = \dfrac{V}{R} = \dfrac{4.5}{9} = 0.5\,\text{A}$ * **1 mark**

> **TIP**
> The current in the series circuit is the same for all the bulbs, but to calculate the current use the total circuit resistance: $3 \times 3.0\,\Omega = 9.0\,\Omega$ and the total e.m.f. in the circuit, 4.5 V.

(ii) $P + I^2 R = 0.5 \times 0.5 \times 3.0 = 0.75\,\text{W}$ for each bulb. * **1 mark**

(b) (i) Resistance of three $3.0\,\Omega$ bulbs in parallel is $1.0\,\Omega$.

Total circuit resistance $= 1.0\,\Omega + 0.5\,\Omega = 1.5\,\Omega$ (The $0.5\,\Omega$ resistor is in series with the parallel bulbs.)

Total circuit current from battery, $I = \dfrac{V}{R} = \dfrac{4.5}{1.5} = 3.0\,\text{A}$ *

Current in each bulb $= 1.0\,\text{A}$. * **2 marks**

TIP

First find the total resistance in the circuit, then the total current supplied by the battery. The circuit current can then be divided up into three equal parts through the three parallel bullbs. The resistance of the three bulbs in parallel could be calculated using the formula but the value can be written down if you stop to think. The equal resistors in parallel make it 3 times easier for the current to flow – so the total resistance is a third of each separate one.

    (ii)   P.d. across each bulb, $V = IR = 1.0 \times 3.0 = 3.0\,\text{V}$. * **1 mark**

(c)   (i)   In circuit A the bulbs are lit, but not to full brightness. * (Applied voltage and current below bulb ratings.) **1 mark**

    (ii)   In circuit B the bulbs are very bright. * (It is usual to operate bulbs at slightly above their rated values to get a very bright light. This, however reduces their working life.) **1 mark**

(d)   Total circuit resistance increases to $2.0\,\Omega$, the total circuit current falls to $2.25\,\text{A}$. However, this current is shared between two bulbs in parallel which get $1.125\,\text{A}$ each. So the two remaining bulbs get even brighter and could burn out. ** **2 marks**

[total 9 marks]

10.(a)   (i) $R = \dfrac{V}{I} = \dfrac{10}{100 \times 10^{-6}} = 1.0 \times 10^{5}\,\Omega$ OR $100\,\text{k}\Omega$ * **1 mark**

    (ii)   p.d. $= 10\,\text{V}$. * **1 mark**

TIP

So far as the steady direct current is concerned, the capacitor has infinite resistance and can be ignored when calculating the circuit resistance. The capacitor is, however, connected directly across the battery and so will be fully charged to the battery voltage of 10 V.

(b)   The capacitor discharges through the microammeter and variable resistor, R. *

The discharge is exponential with the voltage across the capacitor and the current in the circuit decreasing exponentially. * **2 marks**

(c)   (i) $Q = I \times t = 100 \times 10^{-6} \times 10 = 1.0 \times 10^{-3}\,\text{C}$ (or $1\,\text{mC}$.) * **1 mark**

    (ii) $C = \dfrac{Q}{V} = \dfrac{1.0 \times 10^{-3}}{10} = 1.0 \times 10^{-4}\,\text{F}$ * **1 mark**

[total 6 marks]

11. (i) $I = \dfrac{V}{R + r} = \dfrac{12}{4 + 2} = 2\,\text{A}$. * **1 mark**

TIP

Here R is the circuit resistance external to the battery and r is the internal resistance of the battery – the calculation of the circuit current requires the total circuit resistance to be used. The voltmeters read the p.d. across one and both external resistors.

   (ii) Voltmeter 1 reads: $V = IR = 2 \times 1.0 = 2.0\,\text{V}$ *

       Voltmeter 2 reads: $V = IR = 2 \times 4.0 = 8.0\,\text{V}$. * **2 marks**

(iii) The terminal p.d. of the battery is also 8.0 V. *   **1 mark**

(iv) Voltmeter 2 is connected across the two external resistors, but it is also connected directly across the battery terminals. The voltage across the external resistors is exactly what appears across the battery terminals, i.e. 8.0 V **   **2 marks**

> **TIP**
>
> This answer would earn the two marks. However, it leaves the question unanswered as to why the battery terminals do not provide 12 V when that is the stated battery e.m.f. You may, incorrectly, have given as your answer to (iii) either 12 V or 4 V, both of which are wrong. The effect of the internal resistance of the battery is to waste some of the available e.m.f. The p.d. across the internal resistance is $V = IR = 2 \times 2.0 = 4.0$ V. These 4 volts are 'lost' inside the battery, are not available for the external circuit and are deducted from the 12 V e.m.f., leaving only 8.0 V on the battery terminals.

[total 6 marks]

12.(a) (i) $Q = CV = 10 \times 10^{-6} \times 24 = 2.4 \times 10^{-4}$ C. *   **1 mark**

(ii) $W = \frac{1}{2}QV = \frac{1}{2} \times 2.4 \times 10^{-4} \times 24 = 2.9 \times 10^{-3}$ J *   **1 mark**

(iii) $W = QV = 2.4 \times 10^{-4} \times 24 = 5.8 \times 10^{-3}$ J *   **1 mark**

(iv) Half the energy supplied by the battery is converted into heat in the resistance of the circuit connecting the battery to the capacitor. **   **2 marks**

(b) (i) Under these circumstances **charge** is conserved.

Total charge, $Q = 2.4 \times 10^{-4}$ C.

Charge is shared equally so charge on each capacitor is 1.2×10^{-4} C. *   **1 mark**

(ii) Total energy stored $= \frac{1}{2}QV = 0.5 \times 2.4 \times 10^{-4} \times 12 = 1.44 \times 10^{-3}$ J *   **1 mark**

> **TIP**
>
> Connecting two identical capacitors together so that they share charge means that they will have plates of the same charge sign connected together. They are therefore connected in parallel. When connecting capacitors in parallel we add their capacitances. In this case the total capacitance becomes 20×10^{-6} F. Putting the same total charge into a capacitor of twice the size will fill it to only half the voltage, i.e. 12 V.

(iii) The energy stored in the double size capacitor is only half that which was originally stored in the single capacitor. The charge-sharing process involves a halving of the potential difference across the capacitor plates. In effect the charge flows 'downhill' to a lower potential level. Since the energy stored is given by $\frac{1}{2}QV$, for the same total Q, but only half the V, the energy stored will be halved. The current flowing during the charge-sharing process generates heat in the connecting wires which accounts for the 50% loss of stored energy. **   **2 marks**

[total 9 marks]

13.(a)   A sheet of thin card placed across the beam would absorb all the alpha radiation without significantly reducing the beta or gamma radiation. *   **1 mark**

(b)

* straight after leaving field

path of beta radiation
(electrons)

beam of beta
and gamma

* curve – part of a circle
* curve – in correct direction

path of gamma radiation*
(undeflected)

Region of magnetic field
perpendicular to plane of paper
and upwards, out of paper*   **5 marks**

(c)   (i)   No effect on the gamma radiation.

Path of beta particles (electrons) more curved or smaller radius. *   **1 mark**

(ii)   Because the alpha particles have a mass much greater than the electrons (approximately 8000 times greater), they are deflected very little by a magnetic field which would be adequate to separate the beta particles from the gamma radiation. The alpha particles would follow an almost undeviated path and not be separated from the gamma radiation. *   **1 mark**

(d) $F = Bev = 4.0 \times 10^{-4} \times 1.6 \times 10^{-19} \times 2.0 \times 10^{6}$ N *

$= 1.3 \times 10^{-16}$ N. *   **2 marks**

[total 10 marks]

Solutions to Paper 2

1. (a)

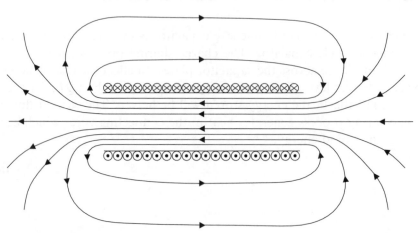

* for correct general shape
* for correct direction of lines
** for equally spaced, parallel straight field lines inside solenoid

4 marks

(b)

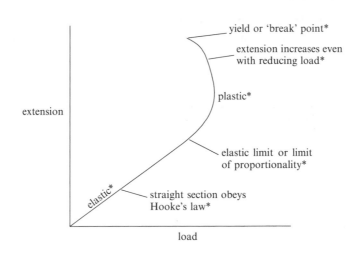

yield or 'break' point*

extension increases even
with reducing load*

plastic*

extension

elastic limit or limit
of proportionality*

elastic*   straight section obeys
Hooke's law*

load

    (i)   * for elastic curve, * for any one key feature   **2 marks**

    (ii) * for plastic curve, * for any one key feature   **2 marks**

                                    [total 8 marks]

2. (a) Output power $= \dfrac{\text{work}}{\text{time}} = \dfrac{\text{force} \times \text{distance}}{\text{time}}$

      Output power $= \dfrac{50 \times 30}{40} = 37.5\,\text{W}$ *

      For 25% efficiency there will need to be four times more power input $(4 \times 25\% = 100\%)$ *
      Therefore the required input electrical power is $4 \times 37.5 = 150\,\text{W}$ *   **3 marks**

TIP

> The output work is calculated from force × distance, where the force required to lift the load is its weight in newtons ($m \times 10$) and the distance is the vertical height. The work can also be calculated using the equation for potential energy gained $= mgh$ which is effectively the same.

  (b) The other 75% of the input power is wasted as it is converted into heat through friction and the heating effect of an electric current in the wires of the motor. *   **1 mark**

                                      [total 4 marks]

3. (a) From the equation for resistance: $R = \dfrac{s \times l}{A}$ where s is the resistivity,

      by rearrangement we get: $l = \dfrac{R \times A}{s}$ $* = \dfrac{20 \times \pi \times (5.0 \times 10^{-5})^2}{4.9 \times 10^{-7}}$ $* = 3.2 \times 10^{-1}\,\text{m}$

      or   0.32 m *   **3 marks**

TIP

> When substituting values in this equation, it is important to be careful to use the correct units. In particular, the units of the radius value (0.05 mm) must be converted into metres: $0.05 \times 10^{-3} = 5.0 \times 10^{-5}\,\text{m}$.
> Remember also, when calculating the cross-sectional area, A, to use the radius value and not the diameter in the formula πr^2.

  (b) The effects of pouring the mercury into a tube of twice the radius are:

      ■ to increase the cross-sectional area of the liquid column by 4 times (radius^2)
      ■ to reduce the length of the column by 4 times. *

      The resistivity of the mercury, of course, does not change.

So the resistance, R, will be decreased by a factor of four as a result of the four-fold area increase and by another factor of four as a result of the reduction of length to a quarter of its previous length.

This results in:

■ a total reduction in the resistance by a factor of 16 *
■ and consequently an increase in the current by a factor of 16 to the value of 1.6 A. *

3 marks

[total 6 marks]

4. (a) $f = \dfrac{c}{\lambda} = \dfrac{3.0 \times 10^8}{400}$ * $= 7.5 \times 10^5 \, \text{Hz}$ *

where c is the speed of light and of radio waves **2 marks**

(b) Photon energy, $E = hf = 6.6 \times 10^{-34} \times 7.5 \times 10^5$ * $= 4.95 \times 10^{-28} \, \text{J}$ *

where h is the Planck constant **2 marks**

(c) Number of photons per second $= \dfrac{\text{total transmitted energy per second (power)} *}{\text{energy of each photon } (hf)}$

Number per second $= \dfrac{400 \times 10^3}{4.95 \times 10^{-28}} = 8.1 \times 10^{32}$ * **2 marks**

[total 6 marks]

5. (a)

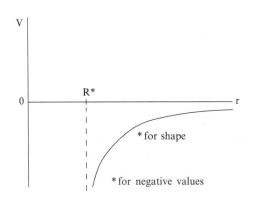

* for shape

* for negative values

3 marks

(b) (i)

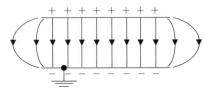

* for shape
* for direction
* for uniform field between plates (i.e. constant magnitude and direction

3 marks

(ii)

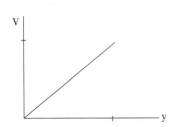

* straight line of posiitve gradient
* goes through the origin

2 marks

[total 8 marks]

6. (a) (i) This is the way in which the average speed of the molecules in a gas is calculated. First the mean value of the squares of all the speeds is found. Squaring the values makes them all positive quantities or ignores all the different directions. This is then square-rooted. * **1 mark**

(ii) The mean kinetic energy of the molecules in a gas depends only on the temperature of the gas. * It is directly proportional to the temperature of the gas in Kelvin. * **2 marks**

(b) (i) momentum of molecule $A = mv = 4.2 \times 10^{-26} \times 600 = 2.52 \times 10^{-23}\,\mathrm{kg\,m\,s^{-1}}$

momentum of molecule $B = mv = 4.2 \times 10^{-26} \times (-400) = -1.68 \times 10^{-23}\,\mathrm{kg\,m\,s^{-1}}$

the mean momentum of A and B is $(2.52 - 1.68) \times 10^{-23} = 8.4 \times 10^{-24}\,\mathrm{kg\,m\,s^{-1}}$ *

1 mark

> **TIP**
>
> Note that because momentum is a vector quantity, the directions of the molecules must be assigned signs. We usually take directions right and up as positive. If you treated momentum as a scalar quantity and ignored the directions of the molecules you would get no marks.

(ii) momentum of molecule $C = mv = 4.2 \times 10^{-26} \times (-200) = -8.4 \times 10^{-24}\,\mathrm{kg\,m\,s^{-1}}$

momentum of molecule $D = mv = 4.2 \times 10^{-26} \times 100 = 4.2 \times 10^{-24}\,\mathrm{kg\,m\,s^{-1}}$

the mean momentum of C and D is $(-8.4 + 4.2) \times 10^{-24} = -4.2 \times 10^{-24}\,\mathrm{kg\,m\,s^{-1}}$ *

1 mark

(iii) mean momentum of all four molecules $= \sqrt{(8.4 \times 10^{-24})^2 + (4.2 \times 10^{-24})^2}$ *

magnitude $= 9.4 \times 10^{-24}\,\mathrm{kg\,m\,s^{-1}}$ *

direction is given by: $\tan\theta = 4.2/8.4 = 0.5$

$\theta = 26°$ down from the direction right. *

3 marks

(iv) mean square speed $= \dfrac{600^2 + 400^2 + 200^2 + 100^2}{4}$ * $= 1.42 \times 10^5\,(\mathrm{m^2\,s^{-2}})$ *

mean kinetic energy $= \tfrac{1}{2}mv^2 = \tfrac{1}{2} \times 4.2 \times 10^{-26} \times 1.42 \times 10^5$ * $= 3.0 \times 10^{-21}\,\mathrm{J}$ *

4 marks

[total 12 marks]

7. (a) Light radiation consists of quanta (or photons) of energy. The energy of each photon is given by $E = hf$, where h is the Planck constant and f is the frequency of the light. It follows that photons of red light which has a lower frequency than blue light will be smaller than photons of blue light. *

The surface of the metal cathode requires at least a minimum amount of energy from an individual photon to be transferred to an electron in the metal surface for it to be released from the surface by photo-electric emission. This is called the work function of the surface. *

The photons of blue light are larger energy quanta than this work function energy requirement, so blue light can release electrons and produce the current registered on the micro-ammeter. The photons of red light have energy quanta which are too small, i.e. less than the work function of the surface, release no electrons from the metal surface and so produce no current in the cell. *   **3 marks**

(b) (i) $E = hf = \dfrac{h \times c}{\lambda} = \dfrac{6.6 \times 10^{-34} \times 3.0 \times 10^8}{4.20 \times 10^{-7}}$ * $= 4.7 \times 10^{-19}\,\mathrm{J}$ *   **2 marks**

(ii) Maximum kinetic energy = photon energy − work function of surface *

$= 4.7 \times 10^{-19} - 2.4 \times 1.6 \times 10^{-19}\,\mathrm{J}$ *

$= 8.6 \times 10^{-20}\,\mathrm{J}$ *   **3 marks**

> **TIP**
>
> To convert the energy of the work function in electron-volts to joules, multiply the voltage (2.40) by the charge of the electron; $e = 1.6 \times 10^{-19}$. Note that the maximum kinetic energy is the left-over energy when the minimum energy required to free the electron from the metal surface (its work function) is subtracted from the incident photon energy acquired by an individual electron.

[total 8 marks]

8. (a) Kinetic energy gained by electrons = work done = $QV = eV$

where e is the charge of the electron.

$eV = 1.6 \times 10^{-19} \times 4.0 \times 10^3 = 6.4 \times 10^{-16}$ J *

Kinetic energy $= \frac{1}{2}mv^2 = 6.4 \times 10^{-16}$ *

$$v = \sqrt{\frac{2 \times 6.4 \times 10^{-16}}{m}} = \sqrt{\frac{2 \times 6.4 \times 10^{-16}}{9.1 \times 10^{-31}}} = 3.75 \times 10^7 \text{ m s}^{-1} *$$ **3 marks**

(b) (i)

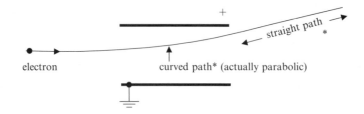

2 marks

(ii) The electric field strength, $E = \dfrac{V}{d} = \dfrac{100}{20 \times 10^{-3}}$ * $= 5.0 \times 10^3$ V m^{-1} * **2 marks**

(iii) The force on each electron, $F = Ee = 5.0 \times 10^3 \times 1.6 \times 10^{-19}$ * $= 8.0 \times 10^{-16}$ N. * **2 marks**

(c) (i)

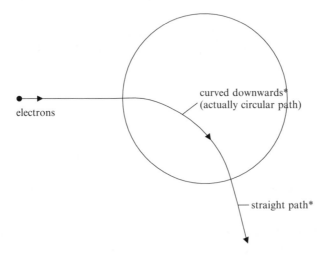

2 marks

(ii) $F = Bev = 4.0 \times 10^{-2} \times 1.6 \times 10^{-19} \times 3.75 \times 10^7$ * $= 2.4 \times 10^{-13}$ N * **2 marks**

(d) (i) In the magnetic field the path of the electrons is an arc of a circle, i.e. it is circular and has a constant radius. *

In the electric field the path is parabolic with the curvature inceasing as the electrons are accelerated towards the upper plate. * **2 marks**

(ii) The circular path which had been flat, i.e. lay in a plane perpendicular to the magnetic field, would become part of a spiral path (still with constant radius) but with a velocity component along the magnetic field direction. * **1 mark**

[total 16 marks]

9. (a) The direction of the induced e.m.f. (or voltage or current) must be such that it will oppose the change which caused it. ** **2 marks**

TIP

Lenz's law of electromagnetic induction can be stated in a variety of ways. This is just one brief version. The law is essentially based on the more fundamental law of energy conservation.

(b)

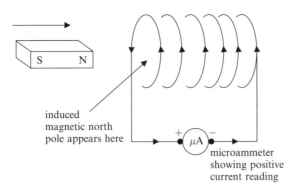

induced
magnetic north
pole appears here

+ μA −
microammeter
showing positive
current reading

** for diagram **2 marks**

A bar magnet is moved quickly towards the open end of a coil of wire (or is pushed into it) *

A current is induced in the circuit of the coil as shown by a microammeter (or any sensitive current or voltage-detecting instrument). There is no source of e.m.f. or current within the circuit itself. *

The direction of the deflection of the ammeter pointer confirms that the direction of the induced current is such that it produces a NORTH magnetic pole on the end of the coil facing the approaching NORTH pole of the magnet. *

Thus the coil tries to oppose the approaching magnet. * **4 marks**

[total 8 marks]

TIP

The fact that the magnet has to be pushed into the coil against a repelling pole induced on the face of the coil requires that work is done in doing the pushing. With a powerful magnet and coil of many turns the effort would be easily felt. The work that is done supplies the energy required to induce a current – with its associated heating effects. So it can be seen that Lenz's law describes an effect which is actually a requirement of the energy conservation law.

---

10. (a) (i) A thermal neutron is one which has been slowed down (by bouncing around between the atoms of the moderator) so that its kinetic energy is comparable with the thermal energy of the vibrating atoms. * **1 mark**

(ii) Neutrons at these slower (thermal) speeds are more readily absorbed by uranium nuclei resulting in further nuclear fissions.

The moderator is required to slow down the neutrons. * **1 mark**

(iii) Deuterium, water or graphite. * It must slow down neutrons without absorbing them. * **2 marks**

(iv) Control rods are required to absorb neutrons, i.e. take them out of the active role of causing uranium nuclei to undergo fission. By absorbing neutrons the chain reaction is slowed down or controlled. * **1 mark**

(v) Boron (steel) or cadmium. * It must be efficient at absorbing neutrons. * **2 marks**

(b) When a uranium nucleus undergoes fission it ejects typically two or three neutrons. These and the larger fission fragments have kinetic energy. *

The source of the kinetic energy is the basic process of conversion of mass into energy which occurs during fission (given by the Einstein equation: $E = mc^2$). *

Kinetic energy of small particles is heat energy. *

As these particles are slowed down, they give their kinetic energy to other atoms * – eventually to those in the cooling fluid – and so the fission energy is carried out of the reactor core as heat. * **5 marks**

[total 12 marks]

11.(a)   Rearranging the equation for a diffraction grating:

The spacing between the lines of the grating, $d = \dfrac{n\lambda}{\sin\theta} = \dfrac{1 \times 6.0 \times 10^{-7}}{\sin 24}$ * $= 1.5 \times 10^{-6}\,\text{m}$ *

2 marks

(b)   First order, $n = 1$: $\sin\theta = \dfrac{n\lambda}{d} = \dfrac{1 \times 4.8 \times 10^{-7}}{1.5 \times 10^{-6}} = 0.32$

$\theta = 19°$ *

Second order, $n = 2$: $\sin\theta = \dfrac{n\lambda}{d} = \dfrac{2 \times 4.8 \times 10^{-7}}{1.5 \times 10^{-6}} = 0.64$

$\theta = 40°$ *

2 marks

(c)   The total number of orders, $n = \dfrac{d\sin 90}{\lambda}$ * $= \dfrac{1.5 \times 10^{-6} \times 1}{4.8 \times 10^{-7}} = 3$ full orders *

2 marks

[total 6 marks]

12.(a)   (i)   The fraction remaining $= \dfrac{1.0 \times 10^{-4}}{6.4 \times 10^{-3}} = \dfrac{1}{64} = \dfrac{1}{2^6} = (\tfrac{1}{2})^6$ *

So there are six half-lives in 42 days.

So the half-life $= 7$ days *

2 marks

(ii)   The decay constant $= \dfrac{\ln 2}{T_{1/2}} = \dfrac{0.693}{7.0}$ * $= 9.9 \times 10^{-2}\,\text{days}^{-1}$ *

2 marks

(b)   ■ Radioactive materials are often stored under water. This acts both as a radiation shield and as a heat-absorbing material. *
■ Some very long half-life materials are vitrified (they are set in solid glass.) This is a very stable medium which will not leak radioactive materials even over hundreds of years. *
■ Liquid materials are sealed in stainless steel containers and protected by very strong outer cases. This prevents leakage and allows safe transport. *
■ Safe storage locations are found. Examples include disused deep mines. Here geological stability of the underground rocks must be considered for long-term storage. *

4 marks

[total 8 marks]

13.(a)   *Any two of the following:*

■ Travelling waves have the same amplitude at all positions along the wave, but the amplitude along stationary waves varies with position between nodes. *
■ There are positions on a stationary wave, called nodes, where, at all times the displacement is zero. There are no nodes on a travelling wave. *
■ Particles in a travelling wave which are exactly one wavelength apart are always in phase. All particles between adjacent nodes in a stationary wave are in phase. *   **2 marks**

(b)   (i)   The fundamental mode of vibration is the one of lowest frequency (and longest wavelength) which can occur in a natural (unforced) vibration of the string. *

1 mark

(ii)   Wavelength $= 2 \times$ length of string in diagram $= 2 \times 120 = 240\,\text{mm}$ *

Maximum amplitude $=$ half the distance between the upper and lower positions of the string at its centre (where there is greatest displacement up and down). *

2 marks

TIP

There are two common errors here. There is a temptation to claim that in its fundamental mode of vibration, as shown in the figure, there is a whole wavelength between the ends of the string. In fact there is only half a wavelength between these adjacent nodes. The upper position of the string is only the crest half of a complete wave. The second common error is to treat the distance between the upper and lower extreme positions of the wave as the amplitude. However, you should remember that the amplitude is always measured from the central or equilibrium position.

(c)

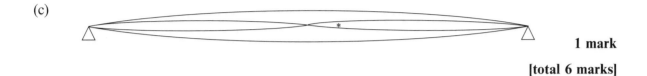

1 mark

[total 6 marks]

14. (a)

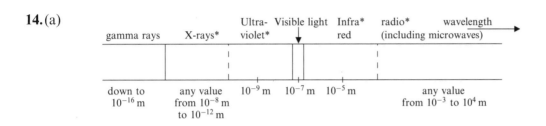

4 marks

(b)   * for each of 4 correctly stated values of wavelength   **4 marks**

(c)   (i)   1. Speed $= 3 \times 10^8 \text{ m s}^{-1}$ in a vacuum. *

2. All are transverse waves. *   **2 marks**

(ii)   1. Magnitude of photon energy $(= hf)$ *

2. Penetrating power, or extent to which it is transmitted through materials. *   **2 marks**

[total 12 marks]

Solutions to Paper 3

1. (a)

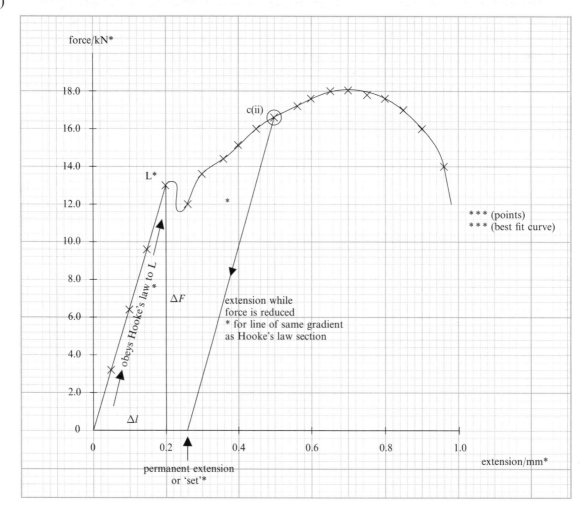

** for correct labelling of the axes of the graph
(the quantity and its units must be correct on each axis)

*** for correct plotting of all the points
(one mark lost for each incorrect point)

*** for best fit curve drawn
(correct shape, good fit to the points, smooth curve, using pencil. Any one of these
qualities missing and a mark is lost) **8 marks**

(b) (i) Hooke's law states that the extension of an elastic material (e.g. spring or steel wire) is
directly proportional to the load applied * providing the elastic limit is not exceeded
(or the limit of proportionality is not exceeded). * **2 marks**

> **TIP**
>
> The most common error here is to omit the reference to the elastic limit. When a spring or
> wire is extended beyond this limit it becomes permanently stretched, but also the shape of the
> graph of extension against load deviated from the straight line of proportionality. It is the
> straight line part of this graph which is said to obey Hooke's law.

(ii) * for correct labelling of part of curve obeying Hooke's law
* for correct labelling of limit of proportionality (L) **2 marks**

(c) (i) The length of the cable would return to its original length, 30.0 mm. *

This is because the stretching had not exceeded the elastic limit. * **2 marks**

(ii)   * for drawing straight line returning to zero load but a permanent extension
* for this line being parallel to the Hooke's law straight section of the extending graph
* for the permanent extension or 'set' being labelled   **3 marks**

(d)   (Data taken from graph)

The Young Modulus $= \dfrac{\text{tensile stress}}{\text{tensile strain}} = \dfrac{F/A}{\Delta l/l} = \dfrac{F}{\Delta l} \times \dfrac{l}{A}^* = \text{gradient} \times \dfrac{l}{A}^*$

where $\Delta l =$ extension and $l =$ original length.

Gradient of graph $= \dfrac{\Delta F}{\Delta l} = \dfrac{13.0 \times 10^3}{0.2 \times 10^{-3}}^* = 6.5 \times 10^7 \, \text{N m}^{-1}$ *

The Young Modulus $= \text{gradient} \times \dfrac{l}{A} = \dfrac{6.5 \times 10^7 \times 3.0 \times 10^{-2}}{1.0 \times 10^{-5}}^* = 1.95 \times 10^{11} \, \text{N m}^{-2}$ *

                                                                                          6 marks

> **TIP**
>
> Watch the units here – a common source of error. The forces are in kN, so $\times 10^3$. The extensions are in mm, so $\times 10^{-3}$. The original length is in mm, so $\times 10^{-3}$: $30.0 \times 10^{-3} = 3.0 \times 10^{-2} \, \text{m}$.

(e)   Energy stored $= \frac{1}{2}$ load $\times$ extension^2 *

Energy stored $= \frac{1}{2} \times 6.4 \times 10^3 \times (1.0 \times 10^{-4})^2$ * (note extension in metres)

         $= 3.2 \times 10^{-5} \, \text{J}.$ *   **3 marks**

(f)   (i)   *Any two valid points.*
The spacing between adjacent atoms is increased slightly along the direction of stretching, * increasing the net force of attraction between molecules. *

The relative positions of the atoms does not change – i.e. there is no slipping of atoms. *
                                                                                          2 marks

(ii)   *Any two valid points.*
Layers of atoms slide against other layers * and this slipping is not reversible. *

The material behaves in a 'plastic' way. *   **2 marks**

                                                                              [total 30 marks]

2. (a)   (i)   The collimator should be adjusted so that it produces a parallel beam of light. *

The width of the single slit should be adjusted so that its images, viewed through the telescope, are bright and sharp (as narrow as is possible while still being bright enough to see clearly). *   **2 marks**

(ii)   The telescope should be adjusted to receive parallel light which it focuses on the cross-wires. * (This is achieved by focusing the telescope on a distant object.)

The eyepiece of the telescope should be adjusted so that a sharp image of the cross-wires can be seen. * (There should be no parallax between the wires of the cross-wires and the image of the collimator slit if the telescope is correctly set up.)   **2 marks**

> **TIP**
>
> The information in brackets is not essential for the 4 marks available, but are relevant points which might gain additional marks on some marking schemes.

(b)

Wavelength/nm	Telescope left°	Telescope right°	mean $\theta°$	$\sin \theta$
448	165.7	194.4	14.35	0.248
501	164.0	196.0	16.0	0.276
588	161.1	198.9	18.9	0.324
668	158.4	201.6	21.6	0.368

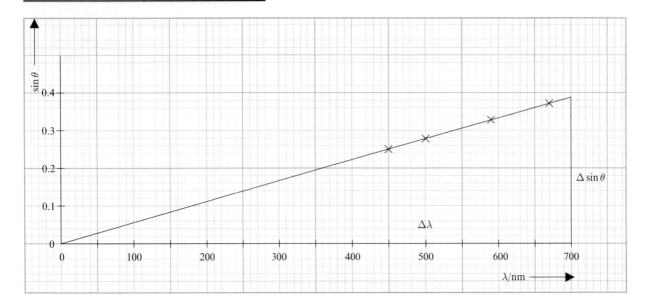

* for axes correctly selected
* for axes correctly labelled with units
** for points correctly plotted (lose 1 mark for each error in plotting)
* for best fit straight line drawn in pencil

Using $n\lambda = d\sin\theta = \dfrac{\sin\theta}{N}$ where $n =$ diffraction order $= 1$,

where $d =$ grating spacing $= 1/N$

and where $N =$ number of lines/m on the grating

gives: $\sin\theta = N\lambda$.

So, a graph of $\sin\theta$ against λ will have a gradient $= N$. *

Gradient $= \dfrac{\Delta\sin\theta}{\Delta\lambda} = \dfrac{0.385\ *}{7.0 \times 10^{-7}} = 5.5 \times 10^5$ lines m^{-1}. *   **8 marks**

(c)   (i)   The diffraction angle $\theta = \dfrac{223.2 - 136.8}{2} = 43.2°$

$d = \dfrac{1}{N} = \dfrac{1}{5.5 \times 10^5} = 1.82 \times 10^{-6}$ m

$\lambda = \dfrac{d\sin\theta}{n} = \dfrac{1.82 \times 10^{-6} \times \sin 43.2}{2} = 6.22 \times 10^{-7}$ m. **   **2 marks**

(ii)   The total number of diffraction maxima is found using the maximum angle of diffraction of 90°.

$n = \dfrac{d \times \sin 90}{\lambda} = \dfrac{1.82 \times 10^{-6} \times 1}{6.22 \times 10^{-7}} = 2$ only ** (third not quite visible)   **2 marks**

[total 16 marks]

3. (a)   Total of all counts $= 5400$

Mean count for 10 minutes $= 540$ *

Mean count-rate per second $= 9.0$ *   **2 marks**

(b)

d/mm	15	30	45	60	75	90	105
N (10 minutes)	1 139 404	495 010	265 802	226 793	115 198	84 608	65 401
Count-rate per second	1 899	825	443	378	192	141	109
Corrected count-rate per second, c	1 890	816	434	269	183	132	100
$\frac{1}{\sqrt{c}}$ $(s^{-1/2})$ *	0.023	0.035	0.048	0.061	0.074	0.087	0.10

(i)   * * * * for data correctly entered in the table (1 mark lost for each error)   **4 marks**

(ii)   * for correct units in bottom left-hand cell   **1 mark**

(c)

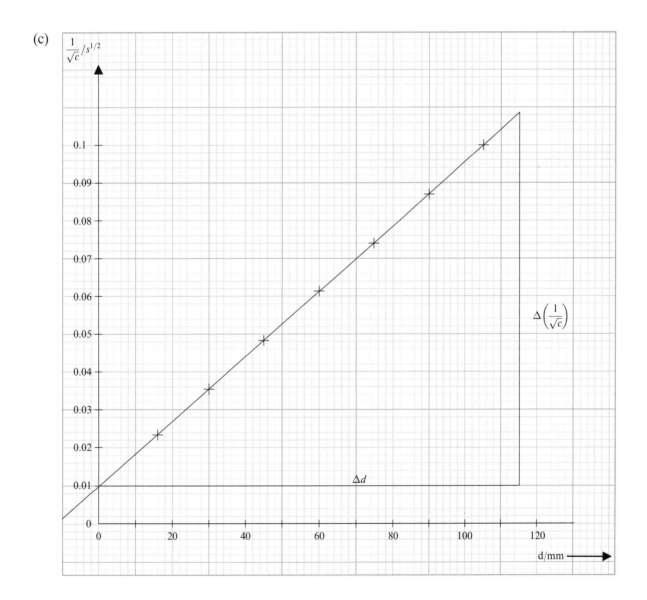

* * for axes and scales chosen sensibly
* * for points plotted correctly (1 mark lost for each error of plotting)
* for best fit straight line drawn in pencil
* for line not drawn through the origin   **6 marks**

(d)   (i)   Gradient $= \dfrac{\Delta\left(\dfrac{1}{\sqrt{c}}\right)}{\Delta d} = \dfrac{0.1\,*}{115 \times 10^{-3}\,*} = 0.87\,\text{s}^{-1/2}\,\text{m}^{-1}$ *   **3 marks**

(ii) The straight line demonstrates direct proportionality between $1/\sqrt{c}$ and d. *

Squaring and inverting, it follows that c is directly proportional to $1/d^2$;

or the count rate is inversely proportional to the distance squared from the source. *

This is evidence of the **inverse-square law** for gamma radiation. *   **3 marks**

(iii) Value of intercept on x-axis $= -10$ mm *

The intercept is the value of d at which c would be infinitely large, i.e. at the true location of the source of the radiation. Thus the intercept is the systematic error in all the distance measurements resulting from not knowing exactly where the source is located. *   **2 marks**

[total 21 marks]

Solutions to Paper 4

1. (a)   Cross-sectional area, $A = \pi \times r^2 = \pi \times (0.75 \times 10^{-3})^2 = 1.77 \times 10^{-6}\,\text{m}^2$. * *

$R = \dfrac{s \times l}{A} = \dfrac{4.20 \times 10^{-3} \times 0.120}{1.77 \times 10^{-6}}\,* = 285\,\Omega$ *   **4 marks**

TIP

It is worth doing the area calculation first and separately to ensure that you remember to halve the diameter and convert the units to metres before squaring.

(b)   (i)   Measurements required:

■ length of specimen between electrical connection points – measured with a mm scale *
■ mean diameter of specimen – mean value obtained from several measurements taken at different points along the conductor using a micrometer screw gauge (reading to .01 mm) *
■ current through specimen read from an ammeter *
■ p.d. across the specimen read from a voltmeter *   **4 marks**

(ii)

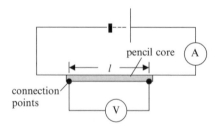

** for circuit diagram drawn and labelled correctly   **2 marks**

(iii) Calculate the mean cross-sectional area (in m^2) from the mean diameter using the formula area, $A = \pi r^2$. *

Calculate the resistance using the formula for resistance, $R = \dfrac{V}{I}$ *

Use the formula for resistivity, $s = \dfrac{R \times A}{l}$ *   **3 marks**

(c)   (i)   A negative temperature coefficient of resistance is one for which the resistance falls as the temperature rises. *   **1 mark**

(ii) A conductor such as graphite behaves like most semiconductor, non-metallic materials. As the temperature rises the dominant factor affecting the conduction properties of the material is an increase in the number of 'free' conduction electrons. The extra electrons are freed by the increased thermal vibrational energy of the latice structure or the atoms of the material. *

More charge-carriers means lower resistance. *   **2 marks**

(d) (i) Power $= I^2R = 0.5^2 \times 100$ * $= 25\,\text{W}$ *   **2 marks**

(ii) 25 W power generated in a small physical space or sample would get very hot to touch. * It could be compared with temperature of an electric light bulb (which is larger) of the same power. * (Any sensible comparison which supports the comment that it would feel hot.)   **2 marks**

[total 20 marks]

2. (a) (i) The natural vibration of a mechanical system is that which results from displacing or disturbing the system and leaving it to vibrate at its own frequency. *

A pendulum (or swing) pushed to one side and released will swing to and fro at its natural frequency. *   **2 marks**

(ii) A forced vibration is one in which an external agency drives a system to vibrate at another frequency determined by the external agent. * The pendulum (or swing) could be pushed back on every return swing earlier than it would naturally return in a free or natural vibration. (This would have the effect of reducing the period of the vibration.) *   **2 marks**

(iii) Resonance is the effect which happens when an external driving agent drives a system to vibrate at (one of) its natural frequency(ies). *

The effect is that the amplitude of the vibration builds up to a large value * as additional energy is fed into the vibration, in synchronisation with it, on each oscillation. *

Timing the pushes of the pendulum (or swing) exactly to match each repeat swing in the same direction causes the amplitude to get larger on every swing. *
                            maximum 3 marks

(b) (i)

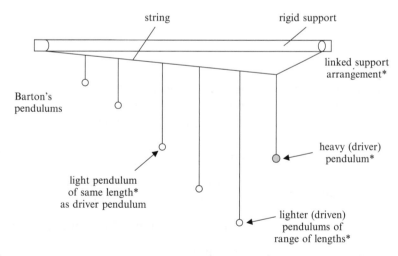

                            4 marks

(ii) The heavy driver pendulum is set swinging so that it feeds energy into the whole system of pendulums. This will start a **natural vibration** of the driver pendulum. *

All the pendulums are set swinging by the rythmic movement (at the driving pendulum's frequency) of the supporting string. This is an example of **forced vibrations**. *

Soon the light driven pendulum (of the same length and natural frequency as the driving pendulum) builds up an amplitude much greater than that of the other driven pendulums. This is an example of **resonance**. *   **3 marks**

[total 14 marks]

3. (a) The efficiency of an electric motor is the fractional (or percentage) value of the output mechanical power * (or work in a certain time) compared with the input electrical power * (or energy in the same time).

OR given as an equation:

$$\text{Efficiency} = \frac{\text{useful output mechanical power}}{\text{input electrical power}} * (\times 100\%)$$ **2 marks**

> **TIP**
>
> Efficiency can be expressed as a fraction (always less than 1) or as a percentage – multiply the fraction by 100 (always less than 100%)

(b) (i)

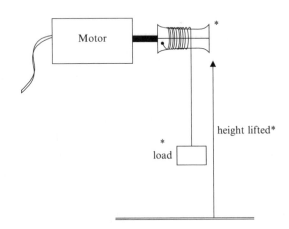

3 marks

(ii)

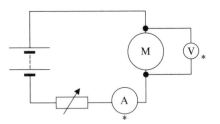

2 marks

(iii) Measurements:

- load lifted in Newtons – use known standard weights *
- height load raised by motor – use metre ruler, value in metres *
- time taken to raise load this height – measured by stop-watch in seconds *
- current input – measured by ammeter *
- voltage across the motor – measured by voltmeter. * **5 marks**

(iv) Output power calculated using: $\text{power} = \dfrac{\text{work}}{\text{time}} = \dfrac{\text{load (N)} \times \text{height (m)}}{\text{time (s)}}$ in watts *

Input power calculated using: electrical power = current (A) × voltage (V) *

Efficiency calculated using equation given in answer (a). **2 marks**

(c) (i) The efficiency is very low because much of the input power is wasted as heat in the wires of the motor and as heat generated by friction in the mechanical movement and because some of the power is required to turn the motor itself. * **1 mark**

(ii) As the load is increased, providing the motor is not so overloaded that it slows down and stalls, the work done in lifting the load becomes a larger fraction of the total work down including work against friction, etc. Thus more of the output is useful power output and the efficiency is higher. * **1 mark**

[total 16 marks]